Gold ore

Gold

Peter Murray

A^+

Smart Apple Media

COPYRIGHT

☼ Published by Smart Apple Media

1980 Lookout Drive, North Mankato, MN 56003

Designed by Rita Marshall

Copyright © 2002 Smart Apple Media. International copyright reserved in all countries. No part of this book may be reproduced in any form without written permission from the publisher.

Printed in the United States of America

☼ Photographs by Tom Myers, Photri (Verna Brainard, R. Harding)

☼ Library of Congress Cataloging-in-Publication Data

Murray, Peter. Gold / by Peter Murray. p. cm. — (From the earth)

Includes index.

☼ ISBN 1-58340-108-3

1. Gold—Juvenile literature. [1. Gold.] I. Title. II. Series.

TN761.6 .M87 2001   669.22—dc21   00-068797

☼ First Edition 9 8 7 6 5 4 3 2 1

Gold

C O N T E N T S

The Gold Rush

One winter day in 1848, James Marshall was digging a canal for John Sutter's new sawmill near Coloma, California. Something shiny caught his eye. He reached into the icy cold water and picked up a bright yellow nugget the size of a pea. There were more nuggets buried in the gravel. Marshall's heart began to pound. *Gold!* The news spread quickly. *Gold! There's gold to be found in California!* ☼ All across America, people quit their jobs, abandoned their farms,

Large gold nuggets

and headed west. The California Gold Rush was on!

Why did people get so excited about a few chunks of yellow metal? Because gold is very rare. But in the early days of the gold rush, gold nuggets and flakes were common in the rivers and streams of California. Gold was found by panning. A prospector scooped some gravel and sand from the stream in a metal pan, swished it around under water, and looked for a glint of yellow. All people needed was a pan and a plan, and they could get rich!

In 1848, a few gold nuggets could buy a horse.

Many prospectors dreamed of instant wealth

Uses of Gold

Since ancient times, humans have gathered gold from rivers and streams. Gold nuggets are so soft that they can be pounded together and formed into many different shapes. Six thousand years ago, the ancient Egyptians used gold coins for money and jewelry. ☀ Gold is one of the softest and most **malleable** metals. It can be drawn into fine wire without breaking. It can be rolled into sheets so thin a person can see light through them. These thin sheets of gold are called

Egyptians used gold leaf to cover masks

gold leaf. Gold leaf is used to gild picture frames and other decorative objects. ☀ Gold is also valued because it is **non-reactive**. That means it does not tarnish or rust like other metals. When treasure hunters find gold coins at the bottom of the ocean, the coins are as shiny as the day they were minted. ☀ Like copper and silver, gold is a good electrical conductor. It is often used for electrical circuits and connectors that require a non-reactive metal. ☀ Dentists use gold for fillings and crowns. The pure gold is mixed with mercury,

Gold the size of your fist could be rolled out to cover a football field.

copper, and palladium to form a hard, durable filling. ⚙

Because pure gold is so soft, it is usually mixed with other met-

als to form **alloys**. Most modern gold jewelry is made from

Ancient gold coins

An open-pit gold and silver mine in Nevada

alloys. An alloy of gold, silver, and copper produces "green gold." Adding more copper makes "red gold." Mixing gold and nickel makes "white gold." These alloys are all extremely strong and long lasting.

Mining for Gold

For thousands of years, people searched for gold in rivers and streams all over the world. Today, most of that easy-to-find gold is gone, but there are still millions of tons of gold in the earth's crust. ☀ Modern gold miners dig deep into the earth to find veins, or deposits, of gold ore. Extracting gold

from the ore is an expensive and difficult process that involves

grinding the rock, mixing it with water and chemicals, and

eventually **smelting** the refined ore in a furnace. Modern

The inside of a gold mine shaft

mining is quite different from the "good old days" when

miners could strike it rich with little more than a pie tin!

The Gold Vault

Gold is as precious now as it was back in the

days of the California Gold Rush. The Fort Knox Bullion

Depository—sometimes called "The Gold **Each gold bar in Fort Knox is brick size and weighs 27.5 pounds (12.5 kg).**

Vault"—in Tennessee contains the largest

supply of gold bullion in the world. Inside

the heavily guarded vault are thousands of

pure gold bars. The gold in Fort Knox is owned by the people

of the United States. ☀ Today, a gold nugget the size of a

jellybean is worth hundreds of dollars. Like our Stone-Age

ancestors, we still use gold for jewelry. A third of all the gold

Pure gold bullion

mined is made into watches, necklaces, rings, and bracelets.

Another third is bought for investments, and the rest is

used in electronics, computer parts, dental

work, coins, and dozens of other items.

For hundreds of centuries, gold has been

valued by humans. A hundred centuries

from now, it might still be one of our most treasured metals.

Some windows have a thin coating of gold. This helps them reflect sunlight and keep buildings cool.

A strip of gold embedded in rock

Panning for "Gold"

What You Need

A metal pie pan   A handful of copper BBs

A cup of dirt (or sand)   A large tub of water

What You Do

1. Mix the BBs with the dirt in the pie pan.

2. Lower the pan carefully into the tub of water. (You should do this outside!)

3. Swirl and shake the pan gently while holding it under the water. The dirt will start to spill over the top of the pan. Keep swirling and shaking the pan.

What You See

Copper BBs are heavier than dirt or sand. The heavy BBs are not lifted as easily as the dirt. This is the same way that prospectors pan for gold. The heavy gold stays in the pan, while the dirt and sand are washed away. If you live near a stream, try panning for real gold. Maybe you will strike it rich!

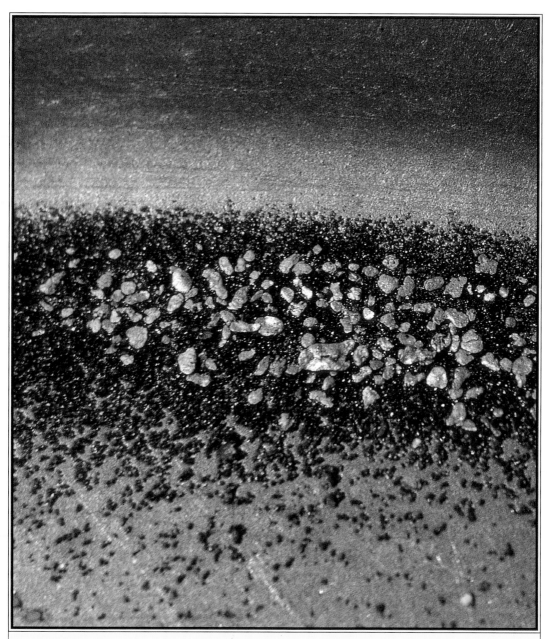

Freshly panned gold nuggets

INFORMATION

Index

Words to Know

alloys (AL-oyz)—mixtures of two metals

gold bullion (GOLD BULL-yun)—bars of pure 24-carat gold

malleable (MAL-ee-uh-bull)—soft and easy to bend and shape

non-reactive (non-re-AK-tiv)—resistant to oxidation (rust or tarnish)

ore (OR)—a mixture of rock and metal

smelting (SMEL-ting)—melting ores to extract metal

Read More

Ito, Tom. *The California Gold Rush.* San Diego, Calif.: Lucent Books, 1997.

Klein, James. *Gold Rush!: The young prospector's guide to striking it rich.* Berkeley, Calif.: Tricycle Press, 1998.

Knapp, Brian. *Copper, Silver and Gold.* Danbury, Conn.: Grolier Educational, 1996.

Symes, R. F. *Rocks and Minerals.* New York: Knopf, 1988.

Internet Sites

The National Mining Hall of Fame

http://www.leadville.com/miningmuseum

The Gold Ledge

http://www.goldledge.com/

The Gold Institute

http://www.goldinstitute.org